A McGRAW-HILL NATURAL SCIENCE PICTURE BOOK

Scientific Adviser: Dr. Gwynne Vevers

Curator of the Aquarium, The Zoological Society of London

Apes and Monkeys

DESMOND MORRIS

ILLUSTRATED BY

BARRY DRISCOLL

McGRAW-HILL BOOK COMPANY

NEW YORK SAN FRANCISCO

MCGRAW-HILL NATURAL SCIENCE PICTURE BOOKS

ANIMALS OF THE ARCTIC by Dr. Gwynne Vevers
Curator of the Aquarium, The Zoological Society of London

ANTS AND TERMITES by Dr. Gwynne Vevers

APES AND MONKEYS by Dr. Desmond Morris
Curator of Mammals, The Zoological Society of London

THE BIG CATS by Dr. Desmond Morris

THE CURIOUS WORLD OF SNAKES by Alfred Leutscher
Co-founder and first Secretary, British Herpetological Society

LIFE IN THE SEA by Dr. Gwynne Vevers

THE STARS by Colin A. Ronan

APES AND MONKEYS

First distribution in the United States of America
by McGraw-Hill Book Company, 1965

First published in Great Britain by The Bodley Head, Ltd., 1964
Library of Congress Catalog Card Number: 65-14378
Second Printing
Printed in Great Britain

Contents

Man's nearest relations

The monkeys and apes are members of a group of animals called the Primates. Man himself is a Primate and the body of a monkey or an ape is similar to ours in many ways. Like us, they can grasp and hold things tightly with their hands. Again, like us, they have rather flat faces with large eyes that are directed forward. Because they can focus both eyes on an object at the same time, they can judge accurately how far away it is. This is important to them when they jump from branch to branch.

One big difference between monkeys and apes is that the monkeys have long tails, which assist them to balance when they are climbing and leaping. Their tails are held out stiffly when they jump and help the animals rather in the way that a circus acrobat is helped by his long balancing pole when he walks the high wire. Apes, like us, have lost their tails during the course of evolution. They, too, have come down out of the trees to live on the ground. There they do not, of course, need to balance themselves so carefully; tails have become unnecessary and have vanished.

Monkeys and apes vary tremendously in size. The tiny Pygmy Marmoset from South America is not much bigger than a mouse and weighs only a few ounces, but the enormous Gorilla from tropical Africa weighs as much as four fully grown men.

The smaller species live for only a few years, but the larger apes have been known to survive for as long as thirty and even forty years in captivity.

Where do they live?

This map shows you where there are monkeys and apes living in the world today. You can see at once that they cannot survive in the colder regions. It is probably the shortage of fruits and nuts for long months that makes it impossible for them to spread further over the continents. Squirrels have managed this because they have developed a system of hoarding their food for the difficult winter days and because they make large, snug dens in which to hibernate. For some reason, no monkeys or apes make big hoards of food, or winter dens, and none of them hibernate. This has restricted them largely to the tropical regions of Africa, Asia and South America.

In South America there are many kinds of monkeys, including the Capuchins, the Howlers, the Woolly Monkeys and the Spider Monkeys. Many kinds of tiny Marmosets and Tamarins are also found in this region. In Africa there are the long-tailed tree monkeys called the Guenons and the Mangabeys, the sturdy ground-living Baboons, the night-living Bush Babies, and the forest-dwelling Chimpanzees and Gorillas. In Madagascar there are the beautiful Lemurs. In Asia, we find the leaf-eating monkeys called the Langurs, the Macaques (Ma-karks) like the well-known Rhesus Monkey, also the acrobatic Gibbons and the huge Orangutan.

The only Primates living wild in Europe are the famous Barbary Apes on the Rock of Gibraltar. They were put there by man, but have survived well in the rather mild climate. Their natural home is in North Africa.

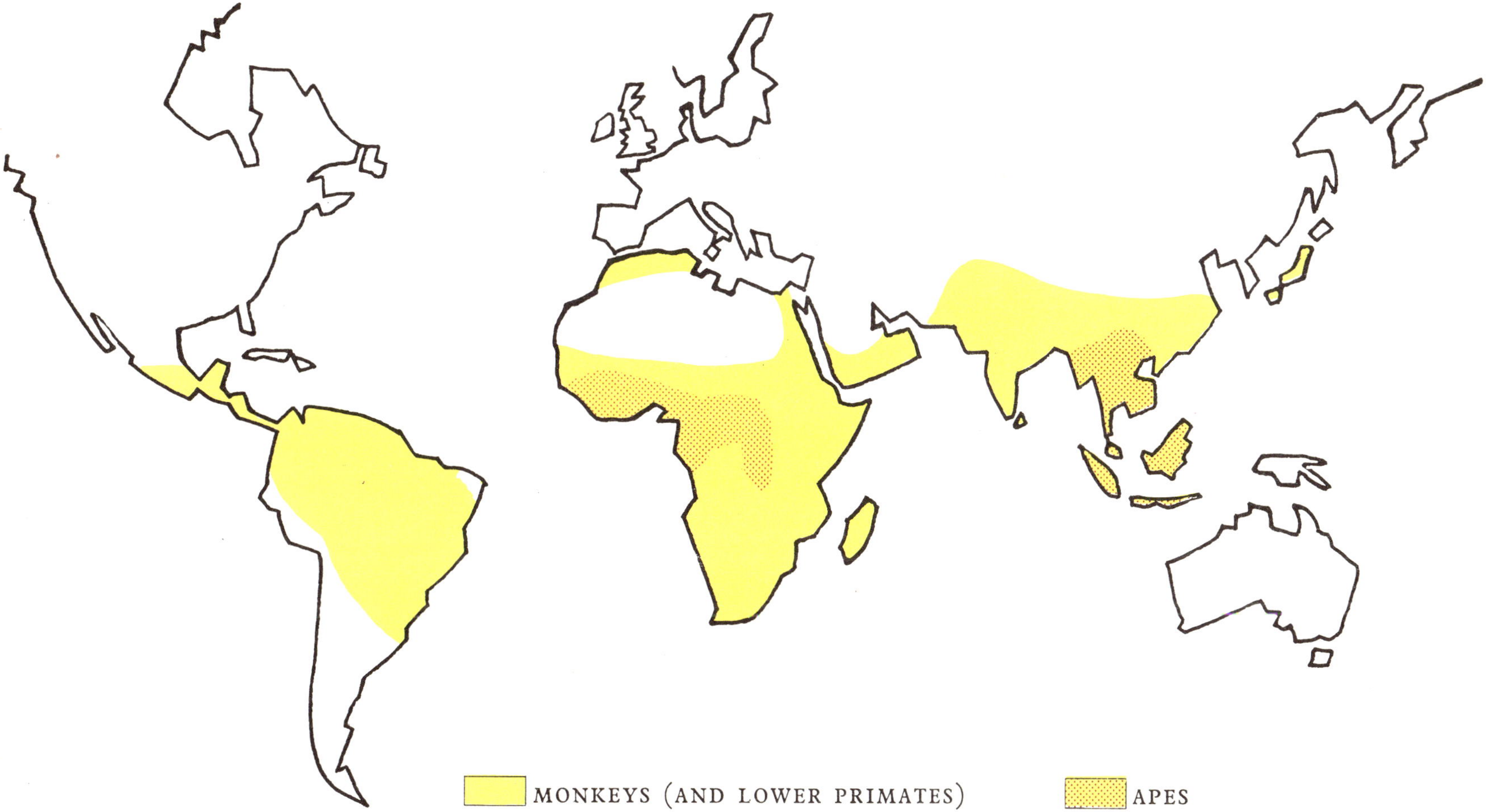
MONKEYS (AND LOWER PRIMATES)
APES

What do they eat?

Monkeys and apes are basically plant-eaters. Most of them prefer to feed on the fruits, berries, or nuts of the trees in which they are living. They often have to vary their diet according to the seasons. At one time, one sort of tree will be in fruit and they will move to that. Then, when the supply vanishes, they will move on to another kind of tree or bush and feed there. Although they never seem to make food stores, some monkeys have cheek pouches which they can fill up with berries or fruits. These pouches are folds of skin that can stretch out on either side of the monkey's face. When it finds a good place, with abundant food, it quickly fills its pouches and then clambers off to a safe spot to eat at leisure.

Some monkeys spend a great deal of time grubbing around on the ground, looking for roots and bulbs. The Crab-Eating Macaque even searches along the edges of rivers and estuaries for any small creatures it can find there. It is not the only monkey to eat small animals occasionally. Many of the little Marmosets and Bush Babies spend hours searching for insects in the bushes and trees. Among the bigger kinds of monkeys and apes, rare cases of meat-eating have been seen. Baboons have been noticed eating snakes, but this is not common. Even the enormous Gorilla is first and foremost a plant-eater. Only man, among the Primates, is a killer and meat-eater on a large scale.

Growing up

The first thing that a baby monkey or a baby ape must do when it is born is to cling tightly onto its mother's fur. It does this instinctively, without any training or learning. It hangs on with both its hands and its feet. Unlike us, it can grip, not only between its thumb and fingers, but also between its big toe and small toes. It stays tightly clasped to its mother's body for some weeks or even months. During this time it goes everywhere with her. She never leaves it in a lair or nest, like so many other kinds of animals. At first, it cannot see clearly, but it searches around in the mother's hair for the nipples where it can feed. This searching, like the clinging, is an inborn action, requiring no learning. The mother will not only feed the baby, but also carefully clean and groom its fur. On cold nights she hugs it closely to her.

After a while, the young monkey or ape will start to venture away from its mother's body. At first, it only goes off for short distances, when she is quiet and relaxed. At the slightest hint of danger, it rushes back to her and clings on again. If it is getting too bold and fails to come back to her, then the mother may dash over and grab it up quickly before there is trouble.

Young Baboons often ride on their mothers' backs rather like jockeys, when they are getting bigger and finding it difficult to hang on underneath her.

Primates of the night

All the larger monkeys and apes are active by day, but some of their smaller relatives only start moving about after dark. They sleep during the daylight hours, curled up in holes or in the forks of trees. Then, as the sun goes down, they wake up and set out on a search for berries, fruit and insects. The insects they hunt are also more active by night, and they have to stalk them through the branches, reaching out swiftly with their hands when they come close to one. These night-living Primates have enormous eyes that are very sensitive even in the dimmest light. The best known is the little Bush Baby from Africa. This is a very agile creature, with long back legs.

In Africa there are also some odd relatives of the Bush Babies called the Pottos. These have shorter legs and heavier bodies and cannot jump at all. They have some sharp spines sticking out of their necks and these can be used as a weapon, if they are attacked and cannot escape fast enough. The spines are part of their backbones and are usually hidden in dense fur, so that when the Potto strikes out with its body the enemy is taken by surprise.

In Asia there are two other kinds of night-living Primates: the Lorises and the extra-ordinary, rat-tailed Tarsiers. You can see a Tarsier catching its prey on the opposite page. In Madagascar there are some night-active Lemurs and in South America there is the strange little "night ape" or Douroucouli (Doo-roo-coo-lee).

The beautiful Lemurs

On the island of Madagascar, off the east coast of Africa, there live some very strange Primates called the Lemurs. They are unlike any of the monkeys found elsewhere in the world. They have tufted ears, rather long snouts and foxy faces that show little expression. Some of them are the most beautifully marked of all the monkeys and apes, having bold black and white patches of fur, or wonderfully ringed tails. The one on the left is the Ring-Tailed Lemur. It often holds its long, bushy tail upright like a black and white flag, but when it is agitated, it crouches down and runs the tail through its arms. It does this over and over again. This is part of the Lemur's "language of smells." On its wrists there are special scent glands, and by pulling the tail over these the animal can cover it with scent. It then acts as a flag that can be be smelled as well as seen.

The strangest of all the Lemurs is a little creature called the Aye-aye. It is so unusual that at first it was classified as a weird sort of squirrel. But careful studies proved it to be a true Primate.

It has very long, slender fingers. Its middle finger is enormously long and with this it taps branches and logs. It listens to the tapping sound and from this it can tell when there is an insect grub beneath the wood. When it finds one, it attacks the wood and scoops the grub out and eats it.

The typical monkeys

Monkeys are very sociable creatures, living together in large groups, or bands. The kinds that live in the forests move about in the trees with ease, leaping, climbing and swinging from branch to branch in search of food. In South America the typical monkeys, like the Capuchins, Howlers, Spider Monkeys and Woolly Monkeys, have a powerful tip to their tails, which can curl around. With this they can grip onto the trees and can even hang from it, leaving their hands and feet free to deal with the problem of grasping for food. In the picture on the right you can see a Woolly Monkey hanging upside down in this way. The tail is so strong that it acts as a fifth limb. For some strange reason, none of the Asiatic or African monkeys (like the one on the left) have developed a tail of this kind.

Many of the African monkeys are beautifully marked with bold areas of black, white, brown and gray fur. Some of the ground-living monkeys, like the Mandrill and the Gelada Baboon, have brightly colored patches of naked skin on the face, chest, or rump. The skin may be red, blue or purple, and these are the most vivid examples of coloring to be found among all the mammals.

The Langurs in Asia and the Colobus Monkeys in Africa are unusual in one important way: they have come to feed largely on leaves. This is not an easy diet for the digestive system and few monkeys have been able to take this step, but those that have managed it have the benefit of abundant food all around them.

The tiny Marmosets

There are many kinds of monkeys in South America, the smallest of which are the Marmosets. They are almost birdlike, with their very high pitched, whistling cries and their brightly marked and colored coats. They are extremely nervous and high-strung, always on the move and always on the lookout for danger. When they start calling out to one another, their twitterings are often supersonic to start with, so that our ears can only hear the last part of their cries.

Unlike the larger monkeys and apes, the Marmosets usually have twins, instead of a single baby. These are so tiny at birth—no bigger than a mouse—that it is often hard to spot them as they cling to the underside of their mother. Marmosets also differ from typical monkeys in having claws, instead of nails, on all their fingers and all their toes except the big toe.

Some Marmosets are pure white, others a beautiful golden or silver color, and still others jet-black. Some have long mustaches and some are adorned with ear plumes. They are so attractive, lively and harmless that they are often kept as pets. But they are very delicate and must be given a great deal of attention and care if they are to settle down and breed well. The pictures show just two of the thirty-three different members of the Marmoset family. On the left is a pair of "Cottontops," with their large, white head plumes. On the right is the Golden Lion Marmoset, a rather rare, brightly colored species. Its whole body is a shimmering, golden yellow.

Why do monkeys make faces?

Like us, many kinds of monkeys and apes can signal to one another by making faces. We know when a man is happy or angry, friendly or miserable, by his facial expressions. We judge this by the shape of his mouth, the position of his lips, the wrinkles on his forehead, the movements of his eyebrows, and so on. Monkeys can do much the same. They too can move their faces into different shapes and these also mean something to other monkeys. One individual can tell another one of its own kind what sort of mood it is in.

A friendly monkey, for example, does what we call lip-smacking: it opens and shuts its lips very fast, making a sort of soft, smacking noise, as if it were tasting something delicious. This shows that it wants to make friends, and if you make this kind of face to a monkey, it may answer you back in the same way. If it is friendly, but also rather frightened of you, it may chatter its teeth at you instead. This means that it would like to come close to you,

but it is not quite sure that you are safe to approach. If you make the lip-smacking face to it, it may then stop teeth-chattering and come near you.

The three pictures show three important monkey faces. On the left is a very angry, aggressive animal, with its lips pushed forward and tightly pursed. Below is a monkey that is angry but rather frightened too. Its lips are slightly pulled back. On the right is a very scared one, with its lips pulled right back and its eyes half-shut.

Having facial expressions of this kind helps to keep the monkeys of a group in touch with one another. It means that they can "talk" to each other with their faces. But the different kinds do not all use exactly the same signals. One must observe each type of monkey separately to learn its special "language" of signals.

Monkeys as pets

Monkeys have been kept as pets since ancient times, but cause all kinds of problems. They are so interesting and attractive that people do not always think of the difficulties when buying one.

Monkeys are especially appealing when they are young and easy to handle. The trouble starts when they begin to grow up. As they become adults, their teeth and jaws get stronger and their bite can be very painful. When they attack someone, they are usually only "defending" their owners and behaving in a perfectly normal monkey way, but friends and relatives that have been bitten are not likely to appreciate this. Eventually pet monkeys have to be shut up in cages for safety. If they have been used to the free run of a room or house, this seems to them like a punishment. At this point they may even turn against their owners. Every week monkey-owners contact the many zoos to ask if they will give a home to pet monkeys that are out of control in this way, but the zoos can only find room for a few of them. So you must think very hard before deciding to get a Primate as a pet.

If you do have one, you should give it a very large cage right from the start, big enough for you to walk into and sit with your pet, like the boy in the picture with his tame Douroucouli. You should provide it with warmth all the year round and a diet of fresh fruit, vegetables, a few insects, some biscuits and vitamins.

The acrobatic Gibbons

In the forests of southern Asia, there live the greatest acrobats of the animal kingdom—the slender, long-limbed Gibbons. Unlike the monkeys, they never run or clamber on all fours. They hang underneath the branches when they leap, swinging from their upstretched arms. Sometimes they will run along a branch for a short way but, when they do, they stand upright, with their arms held high. In the wild they hardly ever leave the trees to come down to the ground. They are completely tailless and are true apes, being smaller cousins of the Great Apes. There are several kinds in different parts of Asia. One kind, called the Siamang, has a large naked pouch of skin on its neck. It can blow this up like a balloon and use it for noise when it is calling to other Siamangs in the forest. Even the other kinds, without pouches to increase the noise of their calls, can make a tremendous din. They give their long whooping cries mostly in the mornings and evenings, and in this way can keep in touch with the others in the trees nearby. Gibbons vary a great deal in color. Some individuals of one kind may be a pale creamy color, others brown, and others completely black. Usually they are found living in small family groups consisting of an adult male and female and their offspring. When very young, a Gibbon clings so tightly to its mother's body that it is often difficult to see as she moves about in the trees.

The Great Apes

There are three kinds of Great Apes living today: the Orangutan in the dense forests of Borneo and Sumatra, in southern Asia, and the Chimpanzee and the Gorilla in the tropical forests of central West Africa. The Orang (far left) has a long shaggy coat of reddish-brown hair, whereas the African apes have black or grayish-black hair. The Chimpanzee (left) is the noisiest and the most active of the three. In the wild it thumps and bangs on trees as it moves through the forest, to keep in touch with the other members of its group. It has a loud voice and uses it frequently, screaming, shouting, or barking, as its mood changes. The Gorilla (right) and the Orangutan are by contrast rather quiet, retiring creatures, and few people have ever been able to study them in their natural homes. It is a tragedy that their numbers in the wild are getting smaller and smaller each year.

The intelligence of apes

The Great Apes are man's nearest living relatives. Unlike him, they have stayed in their original forest homes. They live a simple, fruit-gathering life, moving about in bands from place to place, as the food supplies change. They have not spread into the open country as hunters, as early men did, and they have developed no weapons or complicated language. All the same, they are very intelligent creatures and in captivity they can learn to carry out quite difficult tasks. You often see them performing at the circus, on the stage, or in films. They can learn to ride bicycles, to eat their food with spoons, bowls, and plates, and to drink from cups and mugs. All this certainly shows a high level of intelligence for an animal but in each case the ape had to be carefully trained to make the correct actions. What is more interesting and really much more intelligent is the way a chimpanzee can solve a new problem for itself, without being told how to do it.

If you put a banana into a safe and lock it up with all kinds of bolts, catches, knobs and locks, the ape will fiddle away with these until it has managed to get them all open. Once, a chimpanzee succeeded in unlocking the door of its cage and let all the other apes out into the zoo. It is amazing that an animal can actually learn to select the right key from a big bunch of keys and then put it into its lock and turn it to open it up. Some chimpanzees have even learned to use slot machines at which they can buy their food.

The picture on this page shows what happens if you hang a bunch of grapes from the roof so that it is out of reach and then leave a chimpanzee alone in the room with some small boxes. It will work out how to pile the boxes on top of one another, so that it can climb up and get the reward.

Curious apes

When we say that an animal is curious, we may mean that it is strange or that it is inquisitive. Chimpanzees are both. Their curiosity is endless. They hardly ever seem to sit still. Everything around them is investigated. If they find an object, they pick it up, finger it, shake it, thump it, bite it, and try to open it. In this way they quickly learn all about the world around them. Like us, they are always inventing new games. They will find a rod, or a pole, or a box, or a slope and will soon start to test it out as something to be jumped over, swung around, squeezed under or rolled down. When they are making up new gymnastics like this, they often do so in a special pattern. They start going round and round in a rhythmic way, just as we do in our own gymnastics, or when dancing. If young apes are given pencil and paper, they will transfer this rhythm to the paper, making marks in special patterns like a human infant. They watch what they are doing and seem to be curious to know what is going to appear next. They start to experiment with different kinds of arm movements and find that these make different kinds of patterns on the paper. After a while they are drawing very carefully and controlling the shapes and the position of the lines more accurately. No ape has yet been able to make a drawing of an object we can recognize, but they certainly show great interest in picture-making and will do it for long periods of time without any reward whatever. We have only just begun to study this particular form of chimpanzee intelligence in detail. In this and other ways there is still a great deal to be discovered in the future about the fascinating world of apes and monkeys.

All zoos keep some monkeys or apes, but there are some which have special features of interest to anyone making a study of these Primates, and they are listed below:

U.S.A.

San Diego (Proboscis Monkeys)
Washington (Breeding Gorillas)
New York (Breeding Colobus)
Philadelphia (Breeding Orangs)

CONTINENTAL EUROPE

Basel, Switzerland (Breeding Gorillas)
Antwerp, Belgium (Mountain Gorillas)
Rotterdam, Holland (Breeding Orangs)
Frankfurt, Germany (Pygmy Chimpanzees)
Paris, France (Chimpanzee Island)

ASIA

Peking, China (Snub-Nosed Monkeys)

CANADA

Calgary (Breeding Drills)

GREAT BRITAIN

London (Breeding Orangs)
Whipsnade (Gibbon Island)
Chester (Chimpanzee Colony)
Bristol (Monkey Temple)

SOUTH AFRICA

Pretoria (Breeding Langurs)
Johannesburg (Breeding Baboons)

NEW ZEALAND

Wellington (Breeding Chimpanzees)

AUSTRALIA

Sydney (Breeding Chimpanzees)
Adelaide (Breeding Gibbons)